Collectible Ashtrays

Jan Lindenberger

4880 Lower Valley Road, Atglen, PA 19310 USA

Acknowledgments

A very special thank you to Joyce and Kinneth Rosenthal, from Lakeport, California. Joyce and Kinneth were kind enough to welcome me into their home and allow me to photograph their vast collection of ashtrays. All but a few of the enclosed pieces belong to them as they have been collecting for several years. Thanks again to my dear friends, Joyce and Kinneth. Also thanks to these other contributors: The National Mall in Martinsville, Illinois; and Another Man's Treasures in Milford, Connecticut.

Copyright © 2000 by Jan Lindenberger
Library of Congress Catalog Card Number: 99-64922

All rights reserved. No part of this work may be reproduced or used in any form or by any means—graphic, electronic, or mechanical, including photocopying or information storage and retrieval systems—without written permission from the copyright holder.

Book Design by Anne Davidsen
Typeset in Humanst521/Souvenir Lt

ISBN: 0-7643-0945-5
Printed in China
1 2 3 4

Published by Schiffer Publishing Ltd.
4880 Lower Valley Road
Atglen, PA 19310
Phone: (610) 593-1777;
Fax: (610) 593-2002
e-mail: schifferbk@aol.com
Please visit our website catalog at
www.schifferbooks.com
or write for a free printed catalog.
This book may be purchased
from the publisher.
Please include $3.95 for shipping.

In Europe, Schiffer books
are distributed by
Bushwood Books
6 Marksbury Avenue
Kew Gardens
Surrey TW9 4JF England
Phone: 44 (0)181 392-8585;
Fax: 44 (0)181 392-9876
e-mail: bushwd@aol.com

Please try your bookstore first.

We are interested in hearing from authors with book ideas on related subjects.

Contents

Introduction
4

Chapter One:
Marked California and Numbered Pottery Ashtrays
5

Chapter Two:
Unmarked California and Misc. Pottery Ashtrays
39

Chapter Three:
Blown/Art/Molded Glass Ashtrays
52

Chapter Four:
Foreign Marked Ashtrays
63

Chapter Five:
Popular Glass and Plastic Ashtrays
79

Chapter Six:
Metal/Brass/Copper Ashtrays
97

Chapter Seven:
Figural and Animal Ashtrays
113

Chapter Eight:
Standing and Ethnic Black Ashtrays
132

Chapter Nine:
Advertising Ashtrays
143

Chapter Ten:
Companion Pieces and Lighters
152

Bibliography
160

Introduction

This collectible ashtray information and price guide focuses on pottery ashtrays found primarily in California between the 1930s-1990s. Influenced by Turkish soldiers, smoking became popular among the British and French soldiers after the Crimean War of 1853. By the 1900s, smoking was socially acceptable and even fashionable for men. However, it wasn't until the 1920s that smoking finally became acceptable for women. Ashtrays came in many different styles, shapes, figures, and sizes. They were originally made for functional purposes, but then sold as souvenirs from different states, or promoted certain industries. Many businesses advertised their companies on the ashtrays and used them as giveaways to promote their business. Historical ashtrays like those depicting the Olympics, the World's Fair, famous wars, presidents, centennials, etc., are the most collectible. The pottery ashtrays are also in demand (not just because smoking is being banned in all public places) but because many of the pottery companies are out of business. This smoking item came in many different materials, such as: bronze, porcelain, aluminum, majolica, Depression glass, art glass, wood etc. In the following pages you will find a varied representation of California pottery along with several other famous pottery pieces. We have also included some ceramic and other unusual ashtrays and companion pieces to help make this a more complete price guide. I hope you enjoy this price guide while taking it on your "Smoking hunt," as it is designed to help you identify and price your "finds." Prices may vary according to area found as well as the condition of the piece. Also, prices may differ from shop to flea market to garage sale.

Dates and names of some California potteries are as follows: American Ceramic-1935-1967, Catch Welder-1909-1951, Saschacrostoff-1948-1963, Brayton Laguna-1938-1968, Catalina-1927-1937, Brad Keeler-1939-1953, William Manke-1926-1952, Metlox-1922-1989, Vernon Kilns-1930-1958, Winfield-1929-1962. Note: not all potters marked their pieces. Many pieces just say, USA or California.

Chapter One
Marked California and Numbered Pottery Ashtrays

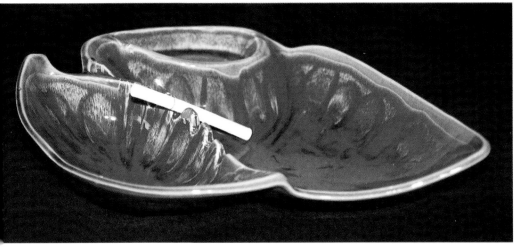

California pottery. Burnt orange and red with brown swirls. 1950s. 12", $25-30.

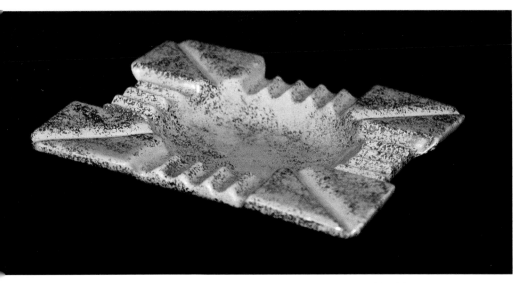

California pottery. Pink with gold flecks. 1950s. 9.5", $25-30.

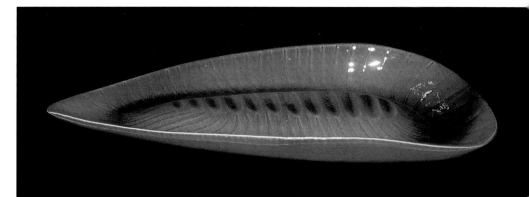

California pottery #A-31. Orange and leaf-shaped with brown swirl. 1960s. 15", $20-25.

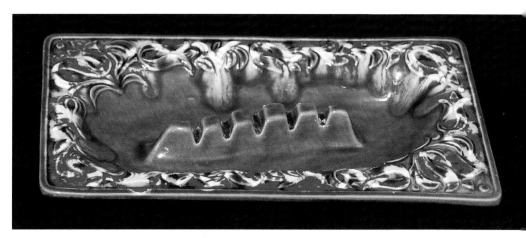

California pottery #1M-105. Oblong, burnt orange with red and brown swirls. Signed Maraca of California. 1960s. 14", $20-25.

California pottery. Pear-shaped, brown with glazed light green inner bowl and white swirled dots. 1960s. 12", $15- 20.

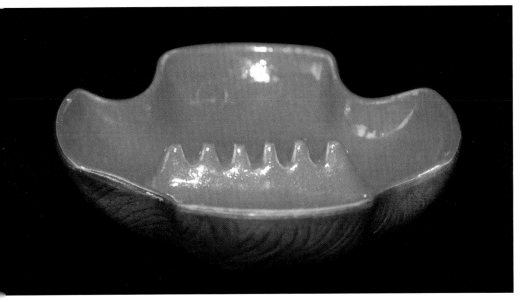

California pottery. Brown with orange glazed inner bowl. 1960s. 8.5", $15-20.

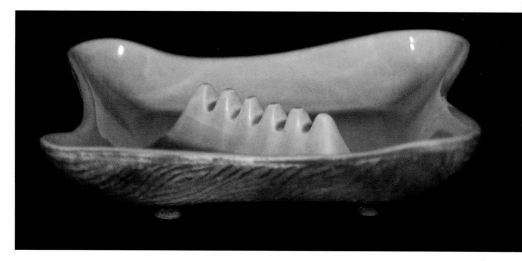

California pottery #1001. Brown with light yellow and orange glazed inner bowl. 1960s. 8" x 8", $15-20.

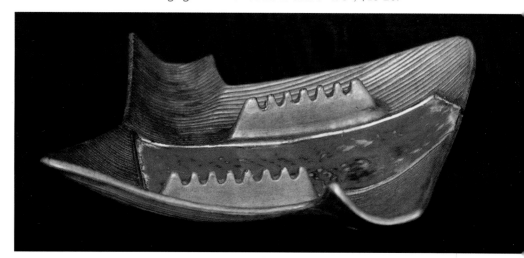

California pottery #118. Brown with orange swirled inner bowl. 12", $20-25.

California pottery #305. Pink and gold swirl with lighter. 1950s-60s. 16", $30-40.

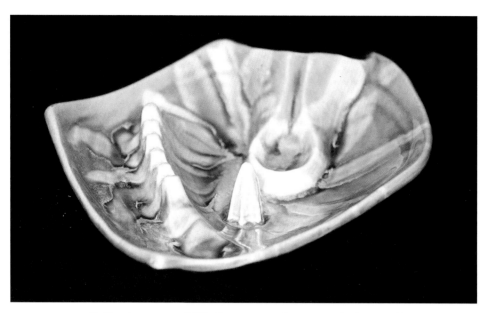

California pottery #691. Pea green with green and white swirl. Maraca of California. 1950s-60s. 12", $15-20.

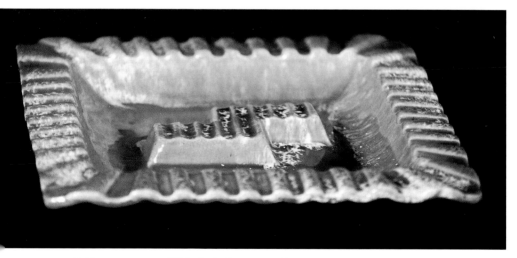

California pottery #332. Light brown and burnt orange with white swirl. Maraca of California. 10", $15-20.

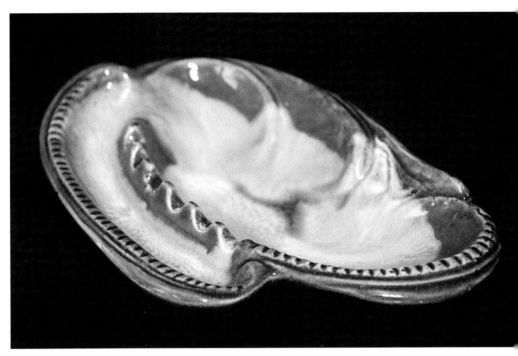

California pottery #678. Dark green with light green center. 1960s-70s. 9", $7-10.

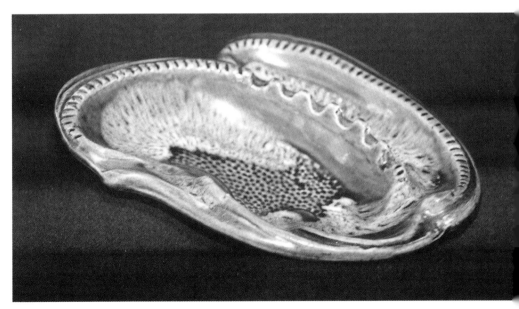

California pottery #678. Two-tone green swirled. 1960s. 9", $7-10.

California pottery- Rosa Originals. Oval-shaped and brown with yellow centers. 5.5", $8-10.

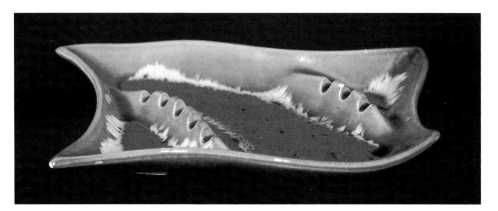

California pottery. Oblong and orange with swirled burnt orange center. 1960s. 8.5", $8-10.

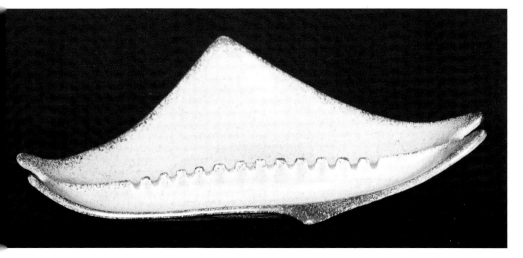

California pottery #371. Star-shaped and aqua with gold speckles. 1950s. 14.5", $20-25.

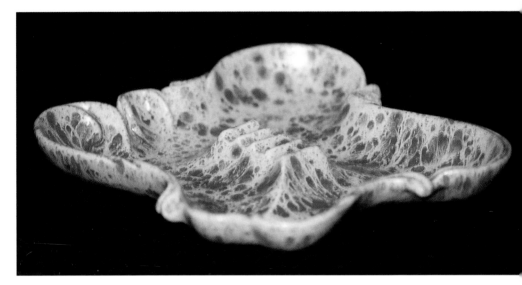

California pottery #P2- Santa Rosa. Butterfly-shaped. Cream with orange speckles and swirls. 7", $12-15.

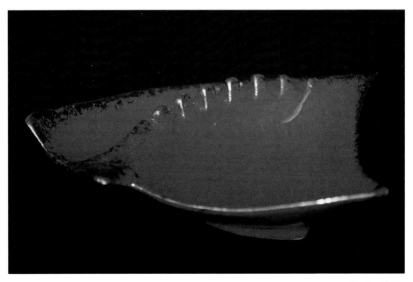

California pottery #614. Orange with gold speckled trim. 1960s. 7.5", $10-15.

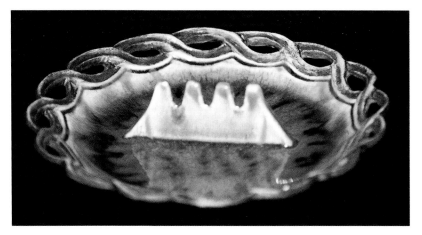

California pottery #767A. Oval, brown and orange with lace edge. 7", $10-15.

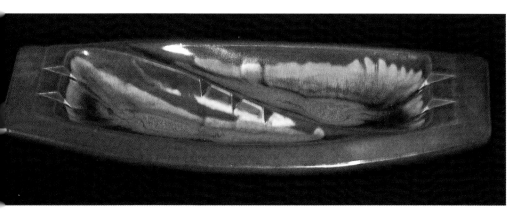

California pottery. F-McF-1961. Oblong and burnt orange with shades of orange swirled into center. 19", $15-20.

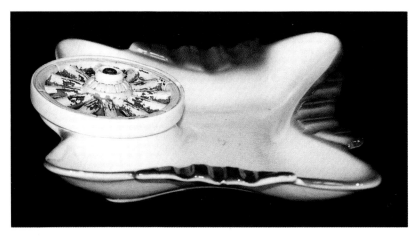

California pottery #250. Square western motif with wheel that turns. Cream with brown edge trim. 9", $15-20.

Unmarked pottery #219. Brown oval with orange speckles. "Shawnee." 6.5", $18-22.

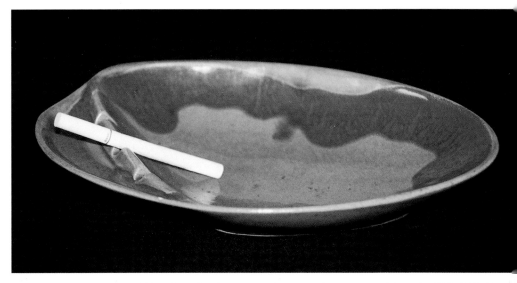

Unmarked pottery #4026- Stangle. Circular and aqua with antique gold swirls. 12", $30-40.

California pottery- Arnell's. Leaf-shaped and burnt orange with green and brown speckles. 10", $10-15.

California pottery #R-138. Oval and pumpkin orange with large, deep orange swirl. 1950s-60s. 10", $15-20.

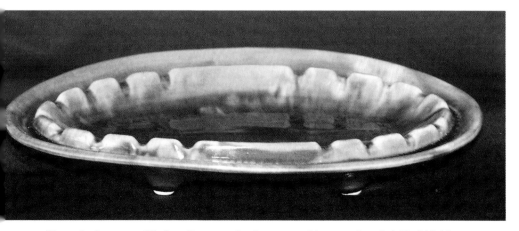

Unmarked pottery- Wades. Ceramic. Oval green and brown glazed. 14", $10-15.

California pottery- CST. Bird-shaped, light brown and cream with glazed aqua inner bowl. 1960s. 6.5", $5-8.

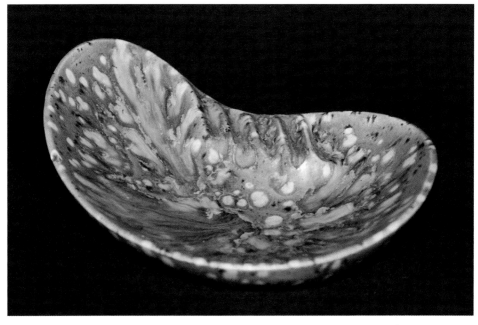

Californai pottery- IOA. Oval green with aqua swirls and speckles. 10", $12-15.

California pottery- Kress. Oval green with aqua trim. 4", $10-15.

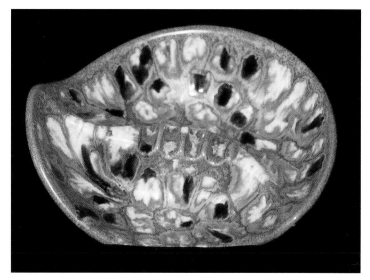

California pottery- Margie's. Oval and brown with white and dark brown swirled speckles. 6", $6-10.

California pottery. Oval and brown with white speckles. 1960s. 11", $12-15.

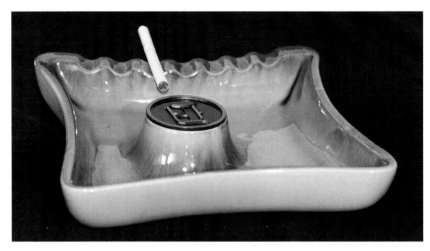

Hull Pottery. The Hyde Park #1990. Aqua square with copper raised center. 8", $40-50.

California pottery #I-04. Long and orange with green, brown and yellow raised design. 8", $15-20.

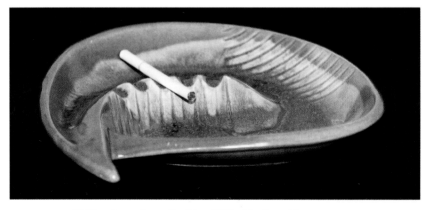

California pottery- Sunkist #671. Black and boomerang-shaped with orange and yellow dots. 10.5", $15-20.

California pottery #212. Oval and tan with green and brown swirls. 1960s. 10.5", $20-25.

California pottery #909. Leaf-shaped and aqua with gold speckles and trim. 10", $10-15.

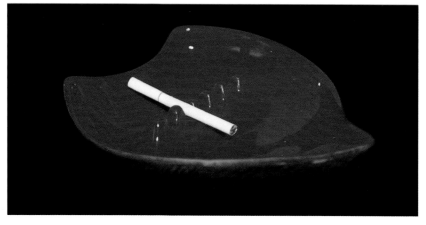

California pottery #7002. Leaf-shaped and brown with red glazed inner bowl. 1950s-60s. 10", $10-15.

California pottery #28. Oblong, gold and white trimmed with green inner bowl. 4" x 9", $8-12.

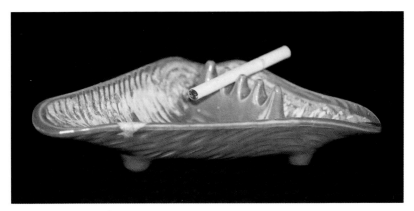

California pottery #151. Triangle-shaped and orange with white swirls. 10", $8-10.

California pottery- USA. Black with raised orange swirls. 10", $8-12.

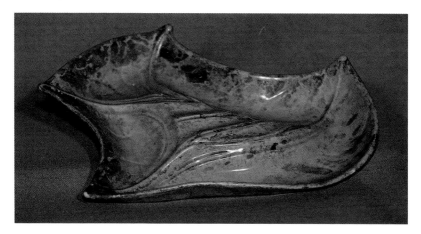

California pottery #L.501. Green and tan raised design. 10", $10-15.

California pottery- USA. Orange and oval with gold speckles. 8", $8-10.

California pottery #0872. Green deco design with a splash of light orange in center. 12", $8-12.

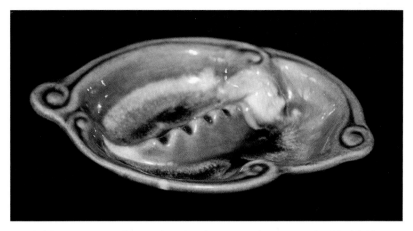

California pottery. Green glazed with aqua and green swirls. 9", $8-10.

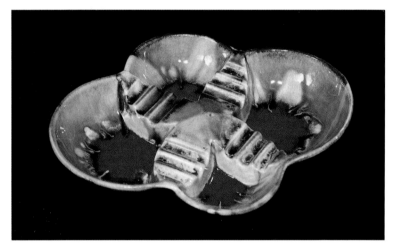

California pottery- USA. Circular, four-sectioned, light orange with deep bowl and red bottoms. 10", $10-12.

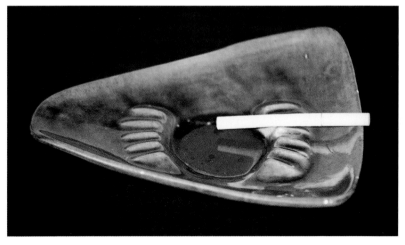

California pottery #D-3. Triangle-shaped and green with orange center. 9", $5-10.

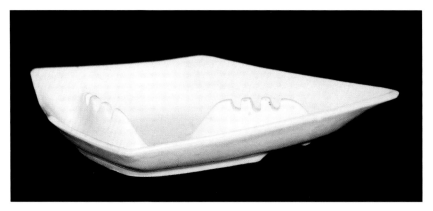

California pottery #2506. Diamond-shaped and white with deep bowl. 13", $20-25.

California pottery #0400. Square and white with gold brushed on trim. 7", $10-15.

California pottery- Winfield, Santa Monica. Set of six white with leaf design in centers. 4", $25-30 set.

California pottery- Laurel of California. Forest green with round matte finish. 7", $20-30.

California pottery #239- USA. Round and brown with green inner bowl. 6", $12-15.

California pottery #699- Wade of California. Shell-shaped, green and brown. 6" and 8", $12-15 each.

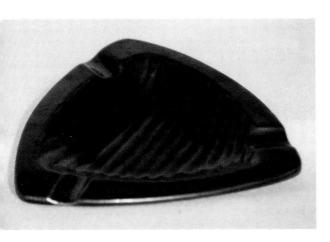

California pottery #202- MR California USA. Triangle. Black with indented bowl. 6", $15-20.

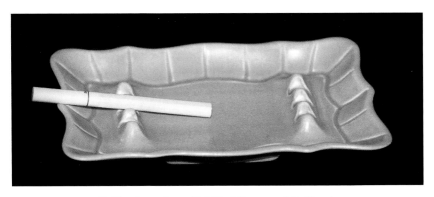

California pottery #136A- Miramar of California. Oblong and yellow with shallow bowl. 7", $10-12.

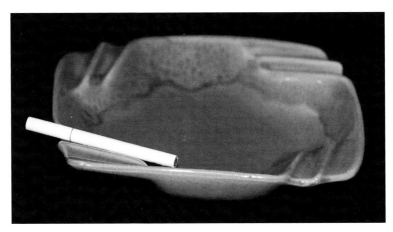

California pottery #3455- USA. Deep orange bowl with shades of orange trim. 8", $15-20.

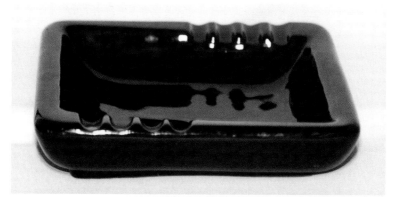

California pottery- USA. Dark green and highly glazed. 9", $25-30.

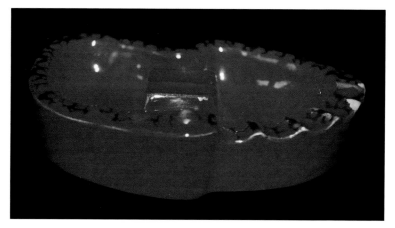

California pottery- Zoe. Red and boat-shaped with gold trim. 7", $10-15.

California pottery #665- Hall USA. Green glazed with raised center. 5", $30-35.

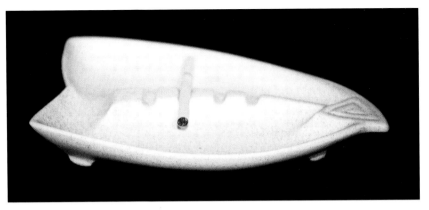

McCoy USA. White speckled with gold feet. 1940s. 9", $30-35.

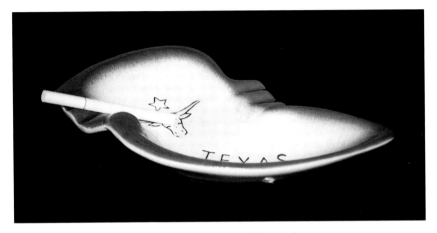

California pottery #T113- USA. White with green trim. Souvenir of Texas. 1960s. 8", $20-25.

California pottery- Maddux #797. Leaf design and light green with gold flecks. 8", $15-20.

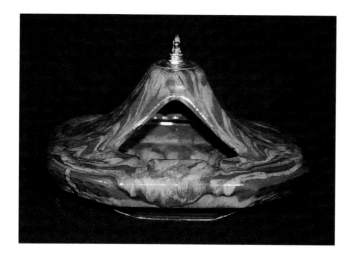

California pottery. Highly glazed hanging ashtray with green and orange swirls. 10", $30-35.

California pottery- Snuff Knob Hill, San Francisco. Round and white with old flecks. Several grooves for cigarettes. 6", $10-15.

California pottery #28- USA. Leaf-shaped and light pink with gold flecks. Removable lid from cigarette holder. 11", $15-20.

California pottery #738- USA. Light pink with brushed gold trim and flecks. 7", $15-20.

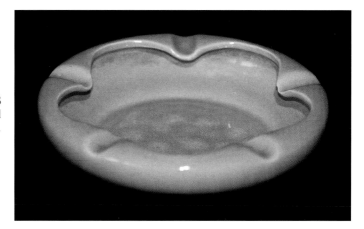

California pottery- GMB USA. Round, deep, and solid pink. 6", $15-20.

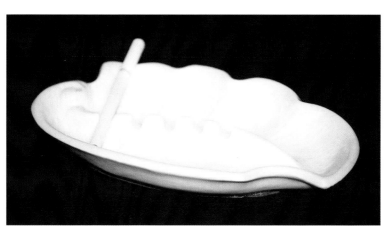

California pottery- Maurice of Ca. Leaf-shaped and buff white. 8", $20-25.

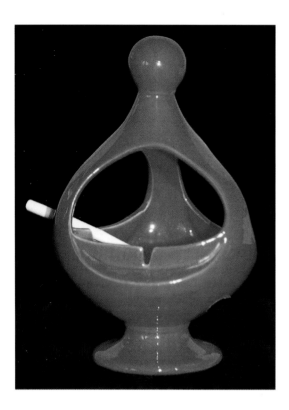

California pottery- Vohann of Capistrano Beach Ca. Standing and sectioned; solid red. 9.5", $20-30.

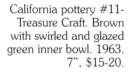

California pottery #11- Treasure Craft. Brown with swirled and glazed green inner bowl. 1963. 7", $15-20.

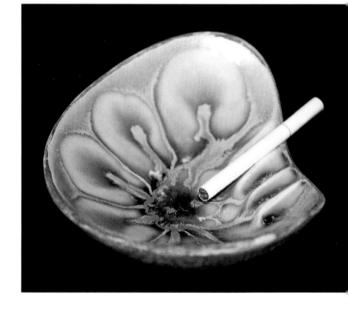

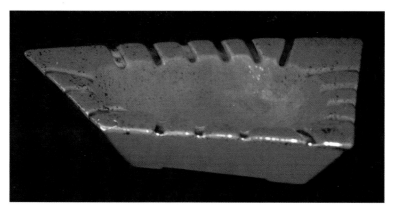

California pottery- USA. Red with gold flecks; odd-shaped. 11", $20-30.

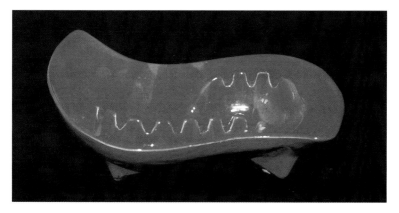

California pottery- USA. Boomerang-shaped with red on legs. 1950s. 8.5", $18-22.

California pottery- Treasure Craft. Brown with green embossed floral, deep inner bowl. 9", $10-15.

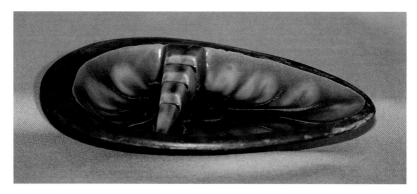

California pottery #10- Treasure Craft. Brown with green highly glazed inner bowl. 1963. 9", $12-15.

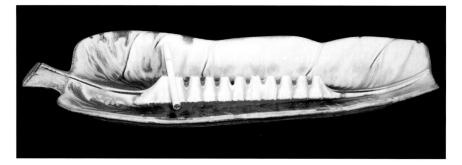

California pottery #23- Treasure Craft. Long, leaf-shaped and orange with cream swirl. 1963. 14", $20-30.

California pottery #22- Treasure Craft. Leaf-shaped and brown with green swirled inner bowl. 1963. 17", $20-25.

California pottery #391- Treasure Craft. Round, cream and green with buff glaze. 6.5", $15-20.

California style pottery #21-67. Crackle glaze and bronze color. $35 45.

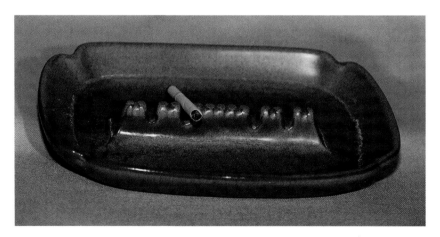

Oklahoma pottery #489- Frankoma. Buffed brown oval. 8", $30-35.

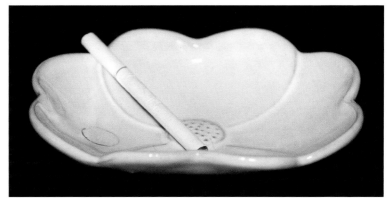

Oklahoma pottery #204- Frankoma. Flower-shaped white buff. 6", $30-35.

Oklahoma pottery #30- Frankoma. Brown and cream swirl bowl. 9.5", $20-40.

California style pottery (east coast)- Heath. Burnt brown with a touch of gold bluff. 6.5", $30-35.

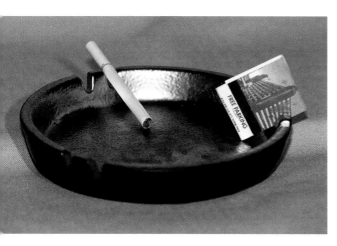

California style pottery- Sasha B. #056A. Gold wash and inner bowl with brown. 7", $35-40.

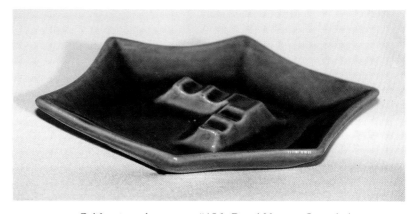

California style pottery #136- Royal Haeger. Six-sided and deep green. 8.5", $20-25.

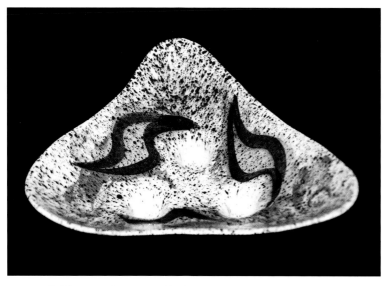

California style pottery- Sasha B. Triangle-shaped with pink and black dots and design. 9", $40-45.

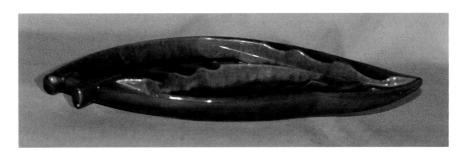

California style pottery #R1156- Royal Haeger.
Deep gray and green leaf- shaped. 18", $30-35.

California style pottery #P125- Royal Haeger.
Aqua with light flecks. 4.5" x 8", $20-25.

California style pottery #224. Large, round and white
with rolled edges and gold trim. 1970. 12", $30-35.

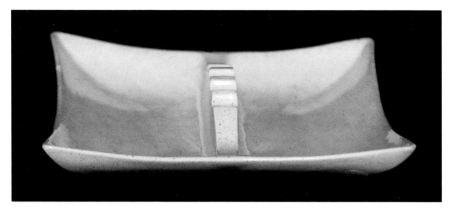

California pottery- Rovano San Francisco. Cream with light gold flecks. 1965. 10.5", $15-20.

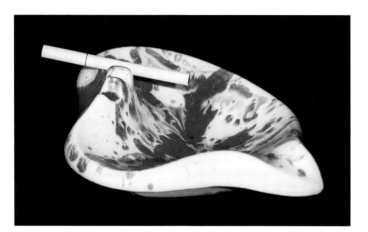

Plymouth Products, Phila., Pa. #75A. Made only one year. White with red swirl. 1959. 7", $15-20.

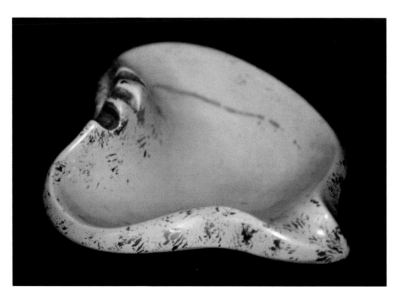

Plymouth Products, Phila., Pa. #75A. Made only one year. Pink, stylized shape with gold markings. 1959. 7", $15-20.

Plymouth Products, Phila., Pa. Made only one year. White with gold flecks. 1959. 8", $15-20.

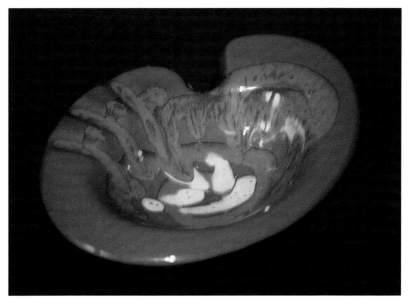

Plymouth Products, Phila., Pa. Made only one year. Green with deep orange and yellow swirl. 1959. 9", $20-30.

Chapter Two
Unmarked California and Misc. Pottery Ashtrays

California pottery- unmarked. Brown and leaf-shaped with turquoise and swirled inner bowl. 7", $10-12.

California pottery- unmarked. Round and burnt orange with black and orange swirl. 8", $10-12.

California pottery- unmarked. Round with gold and soft turquoise. 8", $10-12.

California pottery- unmarked. Oblong, brushed orange and brown. 1960s. 12", $15-20.

California pottery- unmarked. Boat-shaped pink bowl with brown outer layer. 5.5", $8-10.

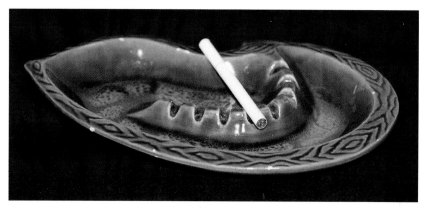

California pottery- unmarked. Light brown and leaf-shaped with green bottom. 10", $10-15.

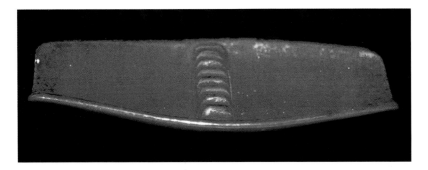

California pottery- unmarked. Long and bright orange. 12", $20-25.

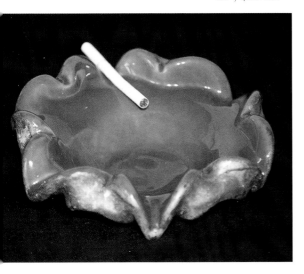

California style pottery- unmarked. Gold paint with a light blue glaze, almost like glass. 8", $10-15.

California pottery- unmarked. Oval-shaped, green, brown and white floral. 12", $20-30.

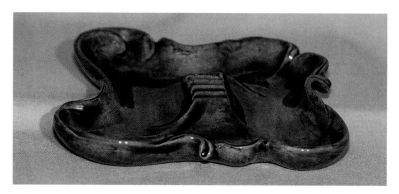

California pottery- unmarked. Green with orange
and brown flecks. 10", $8-12.

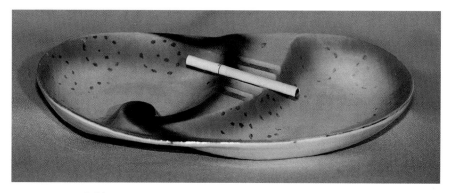

California pottery- unmarked. Oval and white with orange
trim and gold flecks. 12", $15-20.

California pottery- unmarked. White with gold
and green flecks. 1956. 8", $20-25.

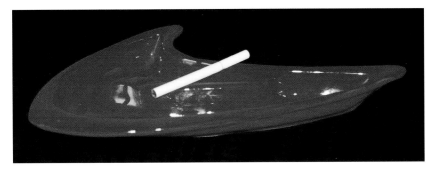

California pottery- unmarked. Boomerang-shaped with red and orange glaze. 8", $15-20.

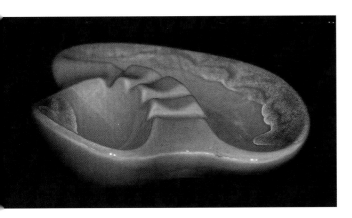

California style pottery- unmarked. Deep pink and orange with green edges. 10", $15-20.

California pottery- unmarked. Green buff with green and aqua flower design. 9", $10-12.

California style pottery- unmarked. Round, embossed design with touches of cream. 11", $15-20.

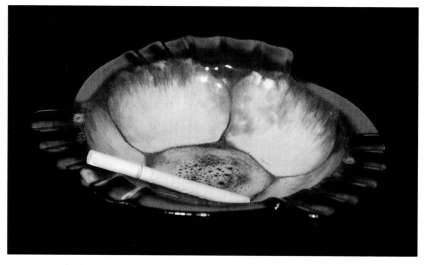

California style pottery- unmarked. Brown with cream swirled bowl and winged edges. 9", $20-25.

California pottery- unmarked. Brown with turquoise brushed edges. 10", $10-15.

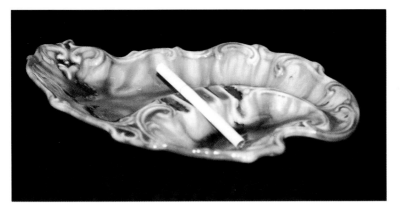

California style pottery- unmarked. Bright yellow with brown and orange swirls. 1960s. 10.5", $20-30.

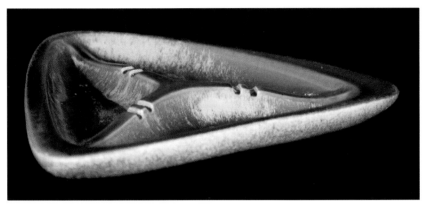

California style pottery- unmarked. Triangle-shaped and white speckled with cobalt blue and white swirled bowl. 15", $20-25

California pottery- unmarked. Round and varied green with a touch of orange and a deep bowl. 6", $8-12.

California style pottery- unmarked. Long and tan with orange and tan swirled bowl. 15.5", $15-20.

California pottery- unmarked. Pink with gold flecks and handle. 3", $6-7 each.

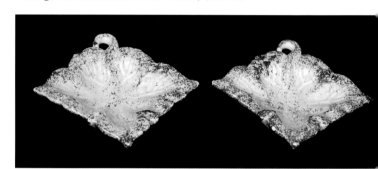

California pottery- unmarked. Oval and green with brown swirls. 6", $8-12.

California pottery- unmarked. Green and circular. 1960s. 9", $7-10.

California pottery- unmarked. Square and light green spotted with deep bowl. 4", $5-8.

California pottery- unmarked. Yellow beaded look with black and red side trim. 7", $20-25.

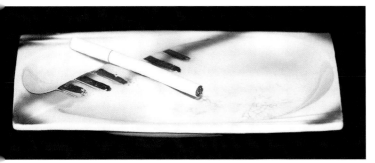

California pottery- unmarked. Oblong and white with gold and aqua trim. 8", $5-8.

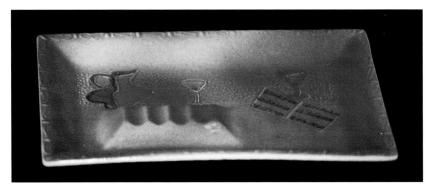

California pottery- unmarked. Pink with cocktail motif. 1970s. 9", $8-12.

California pottery- unmarked. Round and green with match holder. 5", $10-15.

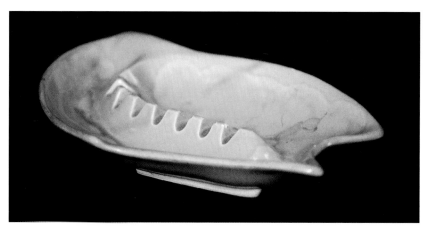

California pottery- unmarked. Yellow with orange swirls. 1960s. 10", $10-15.

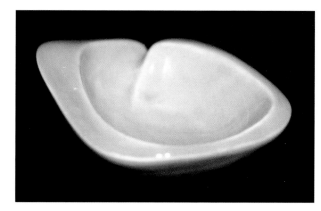

California pottery- unmarked. Circular and aqua with white trim. 3.5", $7-10.

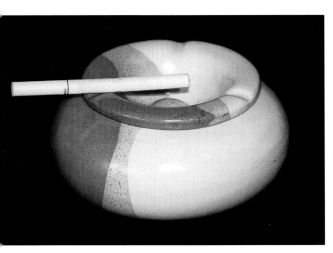

Pottery- unmarked. Ashtray with three tone glazes. 5" x 3", $10-15.

Pottery- unmarked. Triangle-shaped and white with aqua and light brown swirls. 1940s-50s. 9", $10-15.

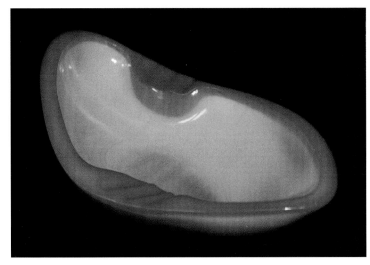

Akro agate. Orange shaded. 10", $40-45.

Pottery- unmarked. White with red inner bowl. 4", $5-8.

Pottery- unmarked. White with yellow inner bowl. 4", $5-8.

Pottery- unmarked. White with aqua and orange inner bowl. 4", $5-8.

Ceramic and round with floral motif. 6", $8-12.

Pottery- unmarked. Round and white with circular orange design. 5", $8-12.

Arizona clay with green inner bowl. Yellowstone Park. 4.5", $8-10.

Chapter Three
Blown/Art/Molded Glass Ashtrays

Blinko art glass. 1950s. 7.5", $50-65.

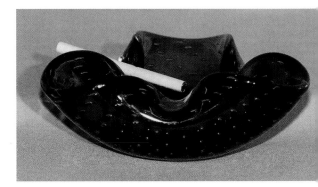

Italian art glass- hand blown. Smoke gray. 7.5", $40- 45.

Italian art glass. Gold and brown. 1950s. 7", $35-45.

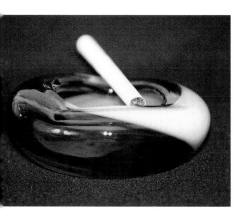

Green and white blown glass.
Unmarked. 4", $8-10.

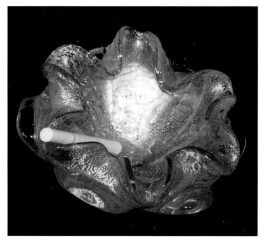

Italian blown glass with gold and silver.
5.5", $40-45.

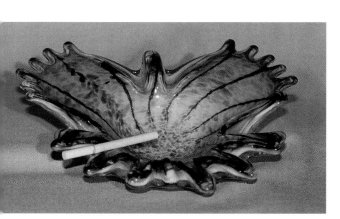

Italian art glass- unmarked.
Green and white with gold
splash. 11", $50-60.

Blown glass-
unmarked.
Peach. 4" x6",
$10-15.

Italian art glass with controlled bubbles. Green. 8", $40-45.

Italian art glass- unmarked. Peach with green trim. 6", $30-35.

Italian art glass. Green. 3 x 7", $45-55.

Italian blown glass.
Graduated bubbles.
Amber. 8", $40-50.

Italian blown glass. Green and clear glass.
6", $40- 45.

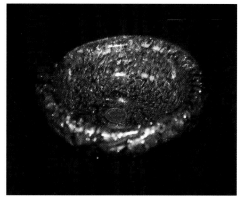

Italian art glass. Murano Italy.
Pink and gold. 4", $15-20.

Art glass- unmarked. Amber with
gold wash. 5.5", $10-15.

Italian art glass. Blue with brilliantly swirled colors. 1950s. 8", $15-20.

Italian art glass. Circular and shades of green. The fish is hollow, laying on top of the base. Unmarked. 7.5", Price unavailable.

Italian art glass. Round with raised sides, shades of light orange color and large bubbles. 8", Price unavailable.

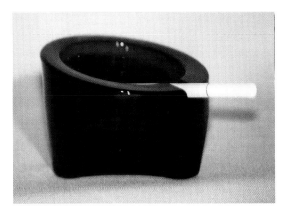

Circular and tall with root beer colored glass.
1940s-50s. 4", $10-15.

Italian art glass. Deep green layered glass. 8", $30- 40.

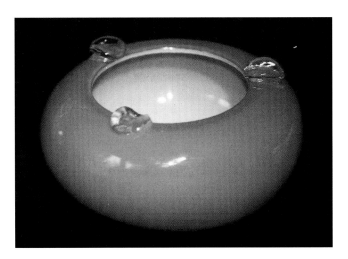

Leftons. Cased and hand blown orange with
white inner bowl. 2.5" x 5.5", $15-20.

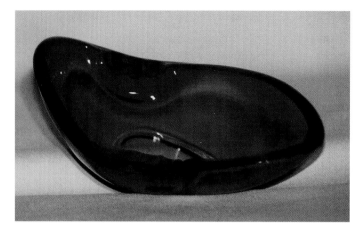

Amberina designer glass. 9.5", $40-45.

Blown circular, clear glass with swirl design in glass. 5.5", $30-35.

Italian blown glass. Orange with bubbles and curved edges. 7", $30-40.

Square blue glass. 4", $10-15.

Blown art glass. Green and deep ashtray with handles. 6", $20-25.

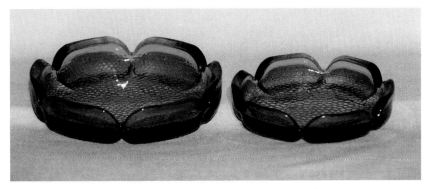

Set of two round and deep aqua ashtrays. 5" & 6", $10-15 each.

Pressed glass. Amber with thumb print design. 1960s. 6", $15-20.

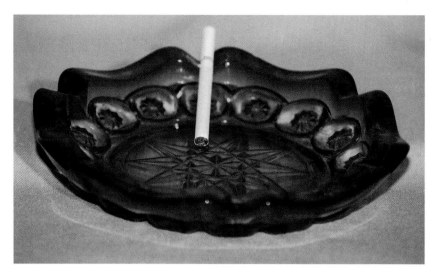

Scalloped design amberina glass. 1950s. 8", $35-45.

Pressed glass. Circular red and orange color with pressed design in base. 10", $40-45.

Amberina color circular glass. Cut block design. 8.5", $40-45.

Circular cobalt scalloped glass. 6.5", $15-20.

Pale amber, nice design glass. 11", $12-15.

Pressed glass. Circular amber with pressed design in base. 10", $40-45.

Light aqua circular glass. 10", $15-20.

Heavy black glass and circular with designed feet. 1950s. 6", $30-40.

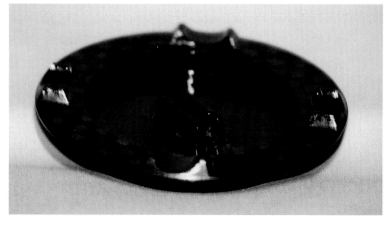

Black circular glass with small, pressed-in block design.
1940s-50s. 6", $15-20.

Chapter Four
Foreign Marked Ashtrays

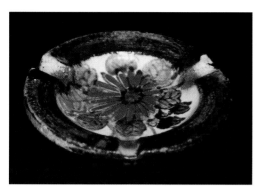

Italian pottery. Brown with floral motif in center. 5", $25-30.

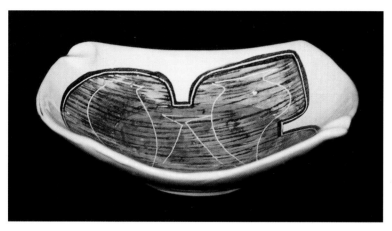

Italian pottery #276/A. Circular and white with blue and orange design in center. 8", $30-40.

Round pottery from Boy's Town, Rome. Green with picture of Boy's Town in center. 6.5", $15-20.

Italian pottery. Square orange and red pin wheel design. 8.5", $15-20.

Italian pottery. Blue nautical design. #00H Italy HH. 8", $20-25.

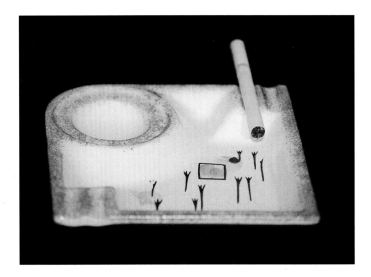

Italian pottery. White square pottery with gold trim and design in center. 5.5", $15-20.

Italian pottery- marked Guillot. Round with yellow edges and butterfly in center. A-La Main, France. 6.5", $25- 30.

Italian pottery- marked Italy. White, and guilded in tan with coat of arms design. 6", $15-20.

Italian pottery- marked Italy. Circular and brown with red pinwheel design in center. 9", $15-20.

Italian pottery- marked Italy. Round and white with gold geometric design in center. 9", $20-25.

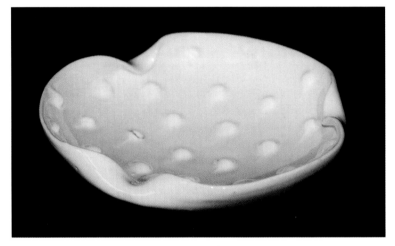

Italian pottery- marked Italy. Handmade oval and white with raised bubbles in bowl. 8", $20-25.

Italian pottery- marked Italy. Leaf-shaped and brown with white and green spotted design. 11", $25-30.

Italian pottery- marked Italy. Turquoise and circular with imprinted design in center. 8", $30-35.

Italian Fior- marked Italia. Restaurant glass ashtray. 4", $8-10.

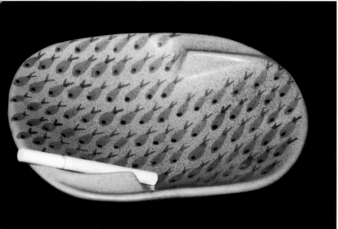

Italian pottery- marked Italy. #190A. Circular with fish design in center. 8", $20-25.

Italian. Enamel over copper, round, signed Cjere. 11", $40-50.

Italian Pottery- marked Italy. Round and white with gray speckling and yellow and aqua stripes. 6", $30-35.

Pottery- marked Japan. Shell-shaped with turquoise tones. 10", $8-10.

Pottery- marked Japan. Boomerang-shaped and bronze. 12", $30-40.

Pottery- made in Japan. Triangle-shaped with orange tones. 8", $6-10.

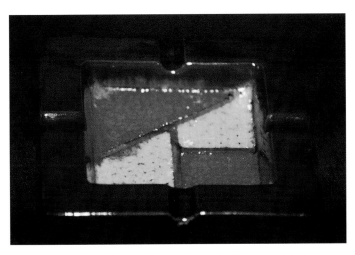

Pottery- made in Japan. Square-shaped with geometric design of orange and yellow. 5", $6-10.

Pottery- marked Japan. White with floral design on front. 3", $10-15.

Teakwood- Japan. Painted burgundy and black with whale painted in center. 4" x 6", $15-20.

Pottery- Japan. Triangle-shaped with dark brown shades. 9", $20-25.

Pottery- made in Japan. White elephants on orange base. 4", $20-25.

Porcelain- Japan. Four piece cigarette and ashtray set. Green with floral design. 6" box, 4.5" trays. $20-30.

Porcelain- Japan. Three white and floral ashtrays with gold trim. 3.5" and 4", $5-8 set.

Pottery- signed Japan. White with floral embossed design. 7", $10-15.

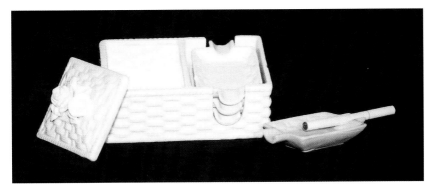

Ceramic cigarette box with ashtrays. Made in Japan. 1940s. 2.5" x 6", $30-40.

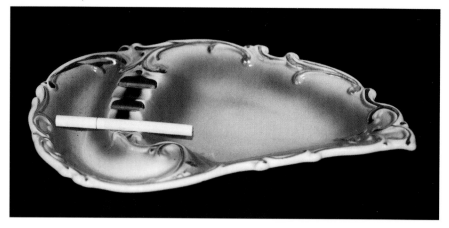

Pottery- Sutton Creations- Japan. Leaf-shaped with pink and gold coloring. 11", $10-15.

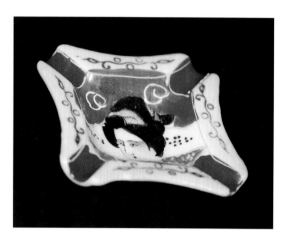

Enamel- made in Japan. Japanese woman in center. 2.5", $10-15.

Pottery- Treasure Craft of Hawaii, Maui. Circular and brown with orange center. 8", $10-12.

Pottery- Treasure Craft of Hawaii, Maui. Oblong Hawaiian motif with browns. 9", $10-15.

Pottery- The Glass House- Hawaii. Oblong shades of brown. 6" x 9", $8-10.

Ceramic- Gaoli's, England. White with Red advertising in center. 5", $8-10.

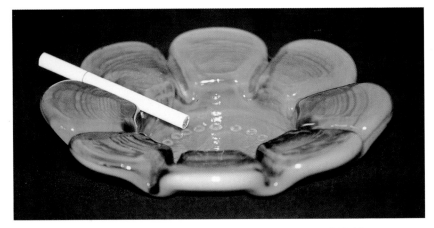

Petal-shaped, heavy blue molded glass. Mexico. 8", $10-15.

Copper handmade coin design.
Mexico. 4.5", $5-10.

Blue and white whimsical glass with squirrel sitting on edge. Mexico. 10", $40-45.

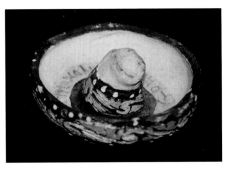

Pottery sombrero souvenir from Grand Canyon National Park. Mexico. 2.5", $5-8.

Wood with glass insert and trimmed in mink. Made in Canada. 4", $20-25.

Pottery- Made in China. White shell design with pipe tray. 8", $10-15.

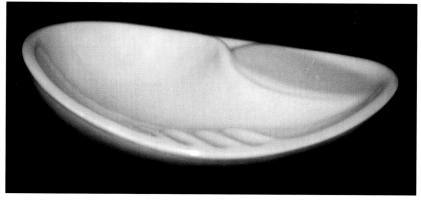
Pottery- signed Veronica Ravano. Oval and turquoise. Alaska. 8/28/64. 9.5", $10-15.

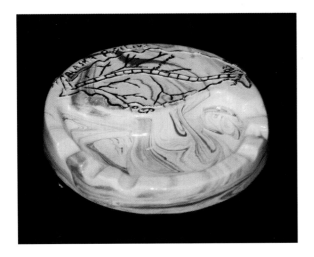

Alaska clay with design of the pipeline. 7", $15-20.

Pottery- signed Bruxelle. White and oval with green house in center. 3.5", $5-8.

Pottery- white with black Arabic design. 3", $20-25.

Pottery- Norway. Handmade pie crust edge with white and yellow design. 6.5", $35-45.

Pottery- signed Hyalyn. Black with red design and white dots. 9", $35-45.

Pottery- made in Norway. #136B-699. Round and white with green and gold edges. 7.5", $25-30.

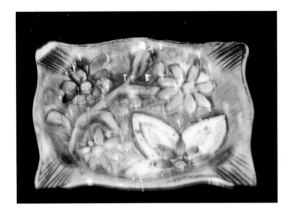

Pottery- Meeqet Quaregnon, Belgium. Light green with embossed flowers in bowl. 2" x 4", $25-35.

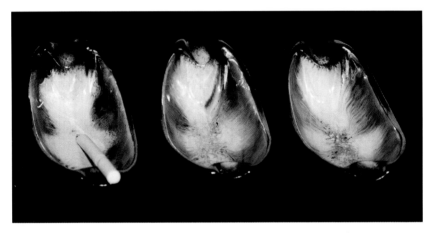

Pottery- Germany, #314. Shades of brown and shell-shaped. 6", $8-10 each.

Pottery- Denmark. Cloc liqueur. 6", $20-25.

Chapter Five
Popular Glass and Plastic Ashtrays

Round and yellow molded glass.
Unmarked. 3", $6-10.

Brown, round glass. Unmarked. 6", $10-12.

Colored and round molded glass. 4", $5-8

Depression Era amber glass. 4", $6-8.

Bar and restaurant ashtrays. Light and dark amber glass. Unmarked. 3.5", $5-7 each.

Amber bubble glass. 8", 6", 4", $20-30 set.

Amber textured glass. 1960s. 4" and 6", $10-15 each.

Triangular, green pattern glass. Unmarked. 7", $10-15.

Square carnival glass. Green and aqua. 1970s. 3.5", $5-7 each.

Triangle-shaped amber glass. 7", $10-15.

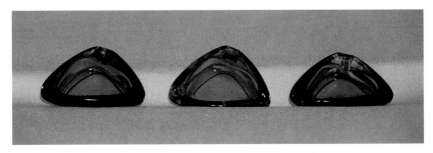

Triangle-shaped. Black, amber, and green. 3.5", $5-7 each.

Different shapes of three ashtrays. Unmarked. 5", $8-10 each.

Molded glass. Round and light gold with marble edge. 5" and 8", $10-20 each.

Clear and red glass with round marble edge. 5", $10-20 each.

Round aqua and square clear with red glass.
Unmarked. 4.5" and 3.5", $6-10 each.

Carnival Depression Era glass. 3.5" and 4", $6-8.

Cranberry Depression Era glass. 4", $10-12 each.

Ruby red and pineapple-shaped glass. 5.5", $10-15.

Round amber and green glass. 4", $6-8 each.

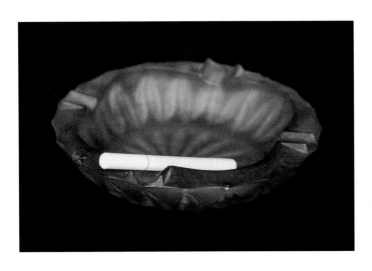

Pink frost with shades of pink and blue. 6", $9-12.

Black Depression Era glass. 3.5", $10-15 each.

Clear pattern glass. 6", $15-20.

Clear pattern glass. Unmarked. 4", $5-7 each.

Round and pressed pattern glass. Unmarked. 7", $10-15.

Round and clear pressed glass with handle. Unmarked. 8", $20-25.

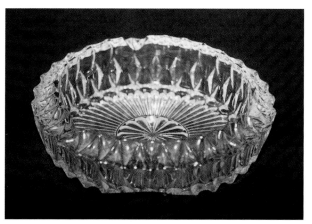

Heavy, pressed clear glass. Fostoria. 7", $25-30.

Clear pressed glass. Italy. 4.5", $20-25 set.

Clear pattern glass. Different styles. 5", $6-8 each.

Left to right: Clear Rosenthal crystal. $20-25. Center: Studio glass. $30-40. Right: Pattern clear glass. $10-15.

Bar and restaurant ashtrays. Clear glass. 3.5", $5-7 each.

Star-shaped light carnival glass. Unmarked. 7", $10-12.

Anchor Hocking clear glass with box. 1940s. $30-40.

Black glass with white string design in bowl.
1970s. 5", $8-12.

Pressed glass, painted red and black. Unmarked. 8.5", $10-15.

Round and white porcelain with green and silver design. 7.5", $8-10.

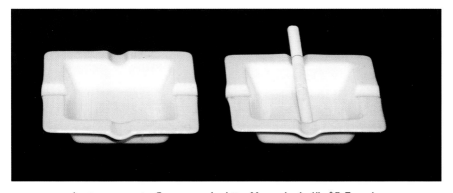

Austere ceramic. Square and white. Unmarked. 4", $5-7 each.

White porcelain with flowers and match holder. Japan. 3", $10-15.

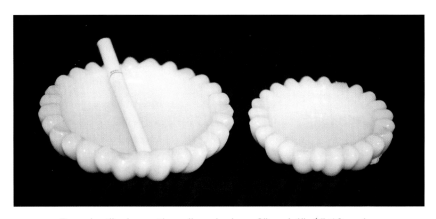

Round milk glass with scalloped edges. 3" and 4", $5-10 each.

Round milk glass with pressed design. Unmarked. 5.5", $8- 10.

Yellow ceramic with floral design. Unmarked. 6", $7-10.

Polished orange alabaster with marble look. Unmarked. 6", $7-10.

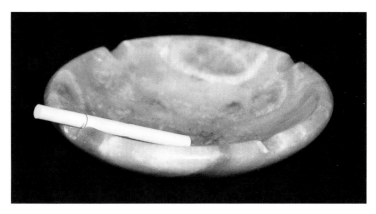

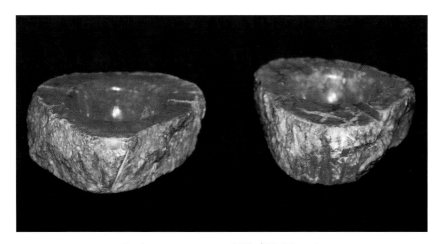

Bright orange quartz. 6.5", $20-30 each.

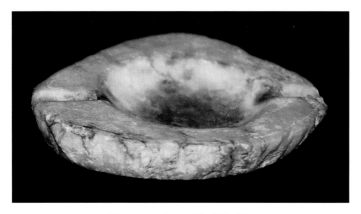

Stone hand cut. 6", $10-15.

Pottery- white and silver tone beaded ring. 6", $20-25.

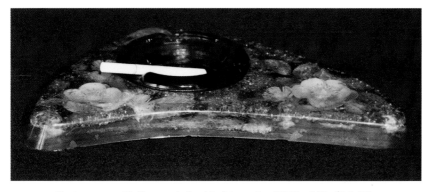

Resin aqua with flowers imbedded in resin. 1960s. 12", $15-20.

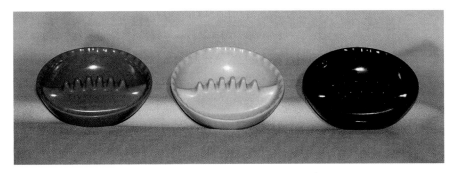

Plastic anholt- green, yellow, black. 1950s. 5", $8- 10 each.

Plastic Anholt- white, yellow, green. 5", $8-10 each.

Plastic- Ges-Line #301, USA. 1950s. 3", $10 each.

Plastic- Irwin-Willert Co. #100, St. Louis, Mo. Blue marble look. 7.5", $20-25.

Plastic- USA. Pre-war. 5", $15-20.

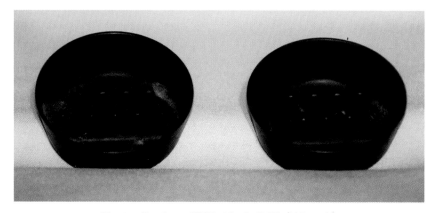

Plastic- Ges-Line #301. Black. 3.5", $10 each.

Plastic- Safti-Plus, #5 MDI Designs. International. Dark green. 6", $10-15.

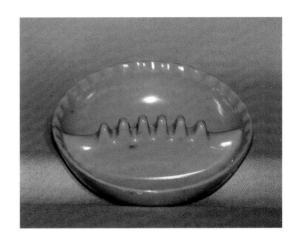

Plastic- Willert Home Products, St. Louis, Mo. Light green. 1950s. 5", $10-15.

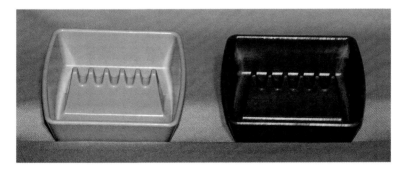

Plastic- Ges-Line #381. Square yellow and green. 4", $10-15 each.

Round bakelite. Willert Home Products Inc. #85-65.
4.5", $10-15 each.

Bakelite- Dale Chemical Co. Yellow and round.
1950s. 4", $10-12.

Chapter Six
Metal/Brass/Copper Ashtrays

Brass and tan swan set. 5", $25-30.

Orange and red cast-iron ladybug. Japan. 1930s. 5", $25-35.

Mahogany wood with ivory on the elephant. Unmarked. 8", $30-35.

Round and royal pewter with brass and enamel turtle.
Singapore. 3.5", $15-20 each.

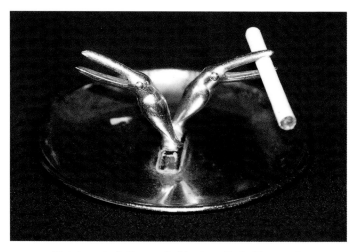

Stainless steel. Toucan holds cigarette in it's mouth.
Late 1940s. 5", $20-30.

Brass and fish-shaped. China. 6", $30-40.

Enamel over copper with fish in bowl. 4.5", $15-20.

Aluminum and horseshoe-shaped.
Unmarked. 4", $10-15.

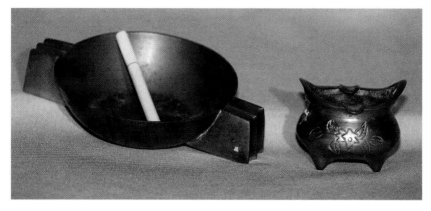

Brass and round with handles. Deco style, 5". Brass pot-shaped with design on front. 2.5", $10-15 each.

Brass with embossed design. Portugal. 7", $20-25.

Brass octagon with embossed design. Marked RWIndia. 6", $20-25.

Copper with detailed design. New Orleans, Natchez. 4", $15-20.

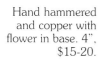

Hand hammered and copper with flower in base. 4", $15-20.

Cast-iron smoke pot. #DE-15.
3", $6-10.

Wrought iron holder and bronze color ashtrays.
5", $20-25.

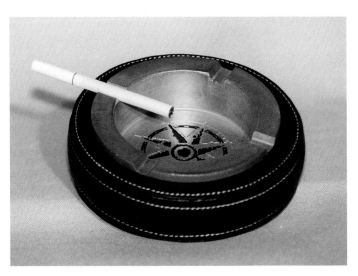

Winston Tire with aluminum insert. Giveaway. 6", $20-25.

Copper and handmade potty.
Plays " You light up my life.
" 8" x 8", $30-35.

Red tin fire bucket. Unmarked.
3.5", $10-15.

Standing copper umbrellas. 5", $10-15 each.

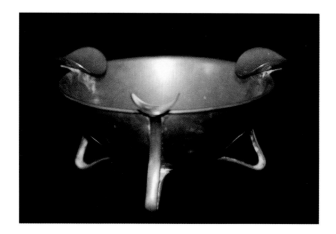

Brass on legs. Unmarked. 2", $10-12.

Copper oriental with detailed design. Japan. 5", $15-20.

Tin with scalloped edge. Early 1950s. $8-10.

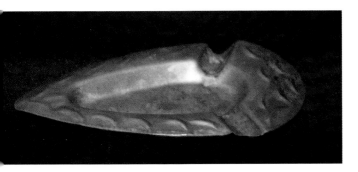

Copper hand with hammered arrowhead shape. 5.5", $12-15.

Oblong. Georgian #514. Hand hammered copper. 4" x 6", $15-20.

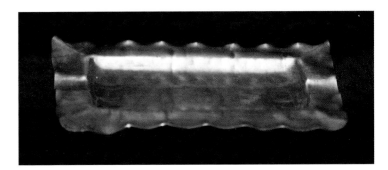

Round stainless steel. Paul Mueller Co.- Springfield, Mo. $15-20.

Round stainless steel. Western Holly, Gas Ranges. 6.5", $20-25.

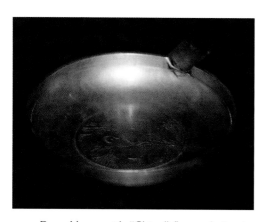

Round brass with "China" flower design in bowl. 3.5", $8- 10.

Brass, Texas-shaped souvenir. Dallas, Texas. 7", $30- 35.

Brass enamel holder and ashtray. China. 4", $20-25.

Round and chrome "Peerless." 5", $15-20.

Polished metal art deco. 2.5", $25-30.

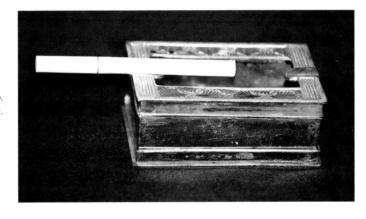

Round and brass. Front: USS Boise, back: CPD. 6", $20-30.

Wood base with aluminum insert. ABG Instrument Engineering Inc. 6.5", $15-20.

Aluminum hand crest, marked with an I. 5", $7-9.

Amber glass insert with wood walnut base. La Salle #11A. 10", $15-20.

Purple aluminum base with glass insert. 1970s. 8", $10-15.

Mosiac tile over aluminum. 4", $10-15.

Round and copper craft guild. Taunton, Mass. 7", $20-25.

Red ceramic tile over copper. Unmarked. 8.5", $20-25.

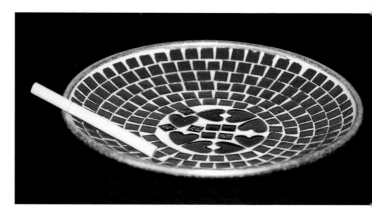

Enamel over copper. Orange and yellow. Unmarked. 6", $10-15.

Beaded set of two with brass corners. China. 4", $10-15.

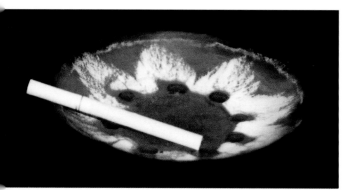

Copper and enamel. Edwards Star Originals. 5.5", $20-25.

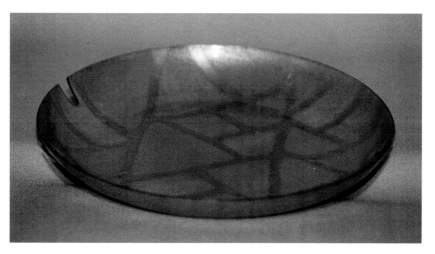

Enamel over copper with geometric pattern. 9", $10-15.

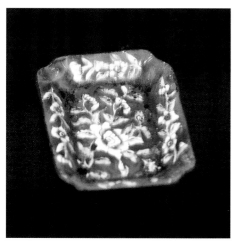

Brass and enamel over copper with floral design. China. 2.5", $10-15.

Orange enamel over metal. 4.5", $3-5.

Copper over enamel with floral design. China. 3", $10-15.

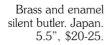

Brass and enamel silent butler. Japan. 5.5", $20-25.

Small copper personal ash catcher. 4", $15-20.

Chapter Seven
Figural and Animal Ashtrays

Terra-cotta. Man sitting on stump. Barrel holds cigarettes. Small basket holds matches. $50-75.

Terra-cotta. Ram being attacked by lion, with cigarette holder. $50-75.

Pottery- black risqué lady. Aloha Hawaii. 4", $20-25.

Pottery #A3772/5- whimsical hat with sleeping man on top. 6", $20-25.

Aluminum whimsical before and after. $35- 45.

Pottery- #1712 Italy. Hand painted Pinochio face. $15-20.

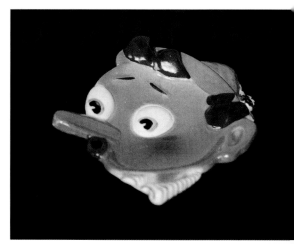

Pottery- jug wrapped in reed. 6", $25-30.

Amber ash bucket. Unmarked. 2.5", $6-8.

Pottery- girl laying on top of bowl. 3.5", $15-20.

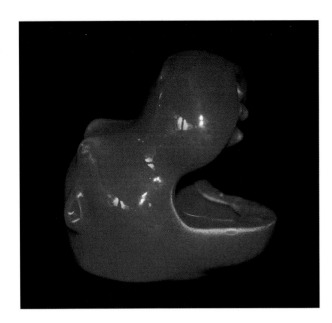

Pottery- orange hippo. Unmarked. $10-15.

Pottery- cream colored elephant. Unmarked. 3.5" x 7", $30-45.

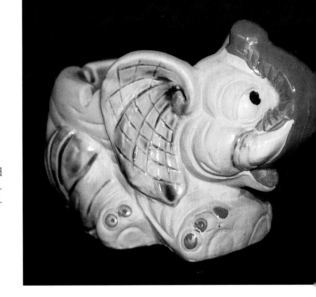

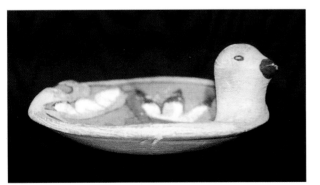

Pottery- bird. Mexico. 5", $12-15.

Pottery- yellow and brown fish. Unmarked. 9", $4-6.

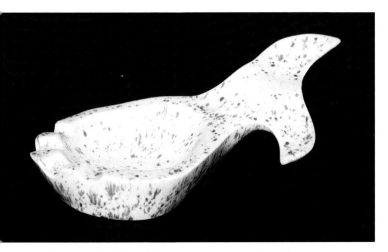

Pottery- White fish with gold and aqua flecks. 10", $8-10.

Pottery- brown pipe with elephant on side. 2" x 4", $10-15.

Ceramic fish with blue and white floral design. Thailand. 5", $8-10.

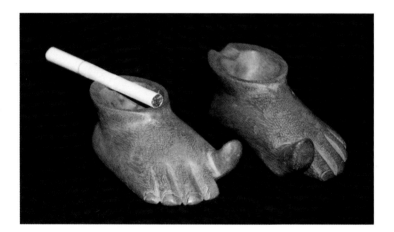

Walnut feet ashtrays. #15-17. 5", $12-18.

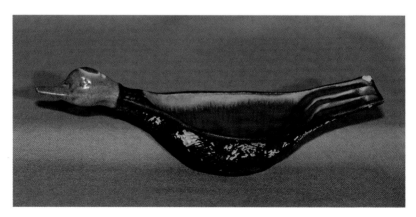

Pottery- brown and orange bowl-shaped roadrunner. California Orig. #5-36. 10", $10-15.

Pottery- white and orange marbled look with moose on top. Handmade by Women of the Moose Lodge. 9", $15-20.

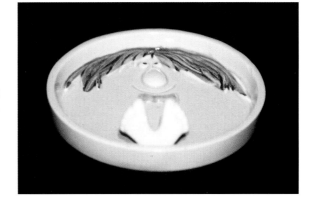

Pottery- round Fritz and Floyd Inc. $15-20.

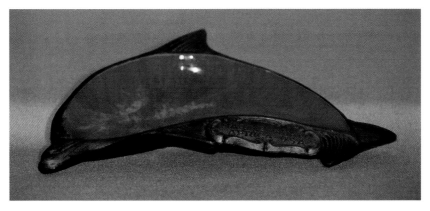

Pottery- Treasure Craft- Made in USA. Brown dolphin with orange bowl. 8", $10-12.

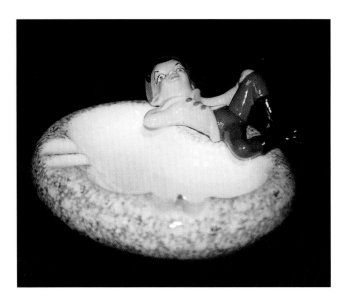

Round pottery with elf laying on side of ashtray. White with spatters of green on sides. 7", $15-20.

Pottery- brown pot belly stove. Unmarked. 5", $8-10.

Pottery- orange and brown pipe. Unmarked. 8", $8-10.

Pottery- brown pipe. Japan. 9", $10-15.

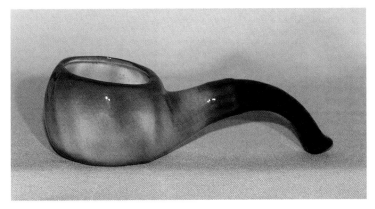

Pottery- brown pipe. Unmarked. 8", $10-15.

Pottery- Napco. Brown pipe. Japan. 9", $10-15.

Pottery- brown and aqua oil lamp shape. 6.5", $10-15.

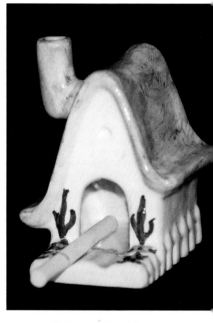

Pottery- little white house. Unmarked. 4", $15-20.

Aqua pottery tee-pee. #51. Arizona. $30-40.

Pottery- yellow bathtub. Souvenir of San Francisco. 4", $10-15.

Pottery- Arnell's. Loggers ashtray. 11", $20-25.

Black pottery with pipe laying on top. E Hall. 10", $15-20.

Jim Beam bottle ashtray. Sold only to members of the club. 1971. $200-300.

Studio pottery. Chimney style stand that holds 4 ashtrays. 4.5", $10-15.

Studio pottery. Chimney that holds 4 ashtrays. Mary Anne. 1976. 5", $10-15.

Pottery- white round scottie on top. 4", $10-15 each.

Glass with metal dog sitting on top. 3", $15-20.

Clear and molded glass bird ashtray. 4", $8-10

Clear glass duck set. Unmarked. 4.5" to 9", $75- 100 set.

Green glass hippo. 7", $10-15.

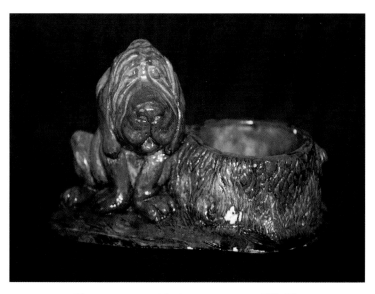

Brown pottery with dog near stump. $20-25.

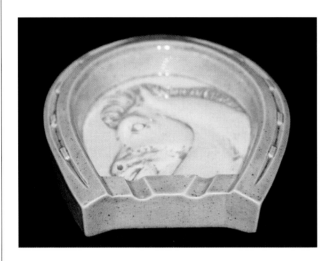

Ceramic horse shoe with horse embossed in center. Unmarked. 6", $10-15.

Blue pressed glass horse shoe with horse in center. Unmarked. 5.5", $10-15.

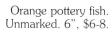
Orange pottery fish. Unmarked. 6", $6-8.

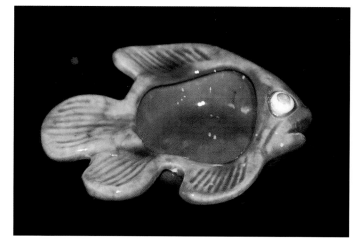

Turquoise glazed pottery fish. 4", $8-10.

Black pottery- Habev. Black glazed bird. 10", $10-15.

Round pottery- Bernard, California. Bird. 7", $10- 15.

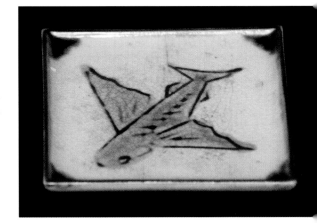

Square, gray pottery with fish in center. Occupied Japan. 5", $15-18.

Square pottery with engraved rooster. Unmarked. 4", $8- 10.

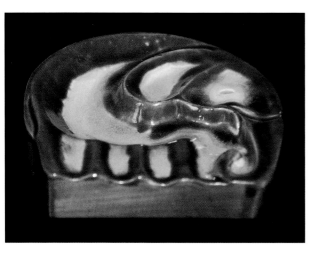

Deep orange pottery elephant. Unmarked. $15-20.

Round leather base with embossed horse head. Dark amber glass insert. $15-20.

Pottery- dark brown and snail-shaped. 6", $20-30.

Chapter Eight
Standing and Ethnic Black Ashtrays

Cast iron standing, stork ashtray. $60-75.

Wooden butler standing ashtray. $60-75.

Tramp art floor ashtray. Late 1800s to 1910. Price unavailable.

Cast iron, pot belly stove replica ashtray with glass insert. 1970s. $60-75.

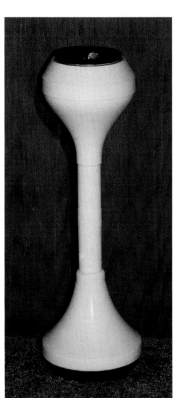

Plastic pedestal floor ashtray. 1960s. $50-60.

Metal stand with pottery ashtray. 1950s. $40-50.

Black iron floor ashtray. 1910-1920. $75-100.

Metal stand that holds glass ashtray. $15-20.

Metal stand with marble ashtray and base. 1940s. $125-150.

Black wooden floor ashtray. 1930s-40s. $50-75.

Iron floor butler with brass ashtray. Reproduction. $100-150.

Brass floor stand with ceramic ashtray and base. 1940s-50s. $50-75.

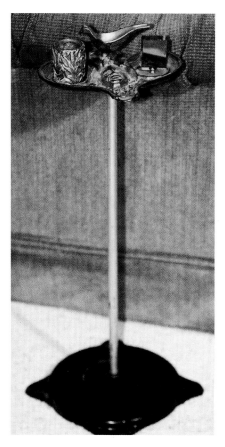

Iron Johnny Griffin standing ashtray. $250-300.

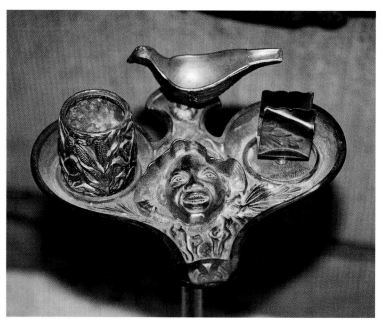

Walnut black butler floor ashtray. $200-300.

Plaster black boy eating melon ashtray. 1950s. $40- 45.

Pottery- native and his hat is the ashtray. 1940s. $50-75.

Plaster native with large lips. Lips are the ashtray. $40-50.

Bisque black boy. His belly is the ashtray and his hands hold cigarettes. 1950s. $50-60.

Plaster mammy with breast in wringer. 1950s. $60-75.

Ceramic boy with donkey and cart. 1950s. $40-50.

Ceramic native holding hat that is the ashtray. 1950s. $40-50.

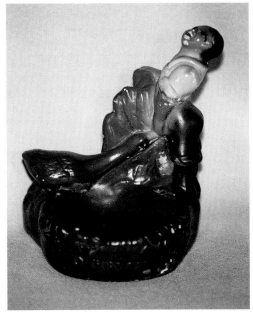

Plaster black boy in sad position with goose. 1940s-50s. $40-50.

Ceramic baby native on hands and knees in bowl. 1930s-40s. $75-100.

Bisque boy playing banjo. He swir back and forth. 1940s. $100-15(

Metal nodding head boy with cigar. 1930s-40s. $150- 175.

Metal nodding head boy with cigar. $150-175.

Ceramic native with shield. 1950s. $40-45.

Metal native with basket on back for cigarettes and front for matches. 1930s. $200-250.

Ceramic native face. 1950s. $40-50.

Ceramic washer woman with four ashtrays in her basket. 1940s-50s. $175-200.

Iron mammy with basket on her head. (Could be a soap dish). 1940s. $200-250.

Majolica black boy leaning on tree stump with cigarette holder. 1930s. $400-500.

Chapter Nine
Advertising Ashtrays

Enesco pottery- yellow and shaped like a pitcher. 8" x 6", $15-20.

Cardboard and metal Chesterfield cigarette ashtray and lighter. 1930s. 2", $40-50.

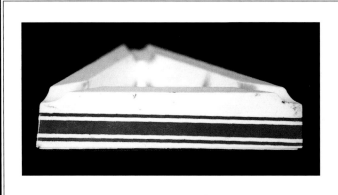

Plastic Camel cigarette advertising ashtray. 6", $10-15.

Half moon and cobalt blue Camel cigarette advertising ashtray. 4" x 8", $20-30.

Diamond-shaped and cobalt blue Camel cigarette advertising ashtray. 4.5" x 7.5", $20-30.

Pewter- CBA. Kansas City, 1977. $15-20.

Round and copper Winston cigarette advertising ashtray. $20-25.

Aluminum Camel cigarette and Burger King advertising ashtrays. $8-12.

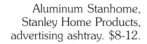

Aluminum Stanhome, Stanley Home Products, advertising ashtray. $8-12.

Clear glass Las Vegas Casinos advertising ashtray with marble edge. $30-35.

Glass casino ashtrays. Harveys and MGM. 4.5", $10- 15 each.

Glass casino ashtrays. Barneys, South Shore Thaoe; Harolds Club, 50th anniversary 1985; Silver spur, Reno. $10- 15 each.

Glass Thunderbird Motor Inns, i.e. Red Lion. 4.5", $5-6.

Clear glass advertising ashtray from Casears Palace, Las Vegas, Nevada. 3.5", $6-7.

Amber glass from Samanski Texaco Service, Cloverdale, Ca. 3.5", $5-7.

Black glass from Nugget Casino, South Tahoe. 3.5", $15-20.

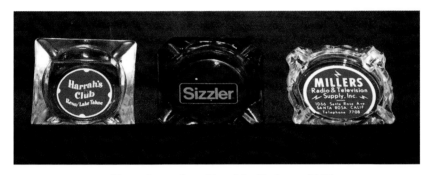

Glass ashtrays from Harrah's, Sizzler, and Millers Radio and TV supply. 4", $10-15 each.

Clear glass advertising ashtrays from the Sheraton Hotel. 5", $15-20.

Clear glass from Excaliber Hotel and Casino, Las Vegas, Nevada. 4", $10-15.

Gold colored glass with painting on back side. Blue Mountains NSW. 5", $6-8.

Ceramic and white advertising ashtray from Hotel. D. Fes. $10-15.

Pottery- made in Italy. Advertising ashtray for
Asti Spumanti wine. 5", $10-15.

Firestone tire advertising ashtray with box.
7", $100-145.

Chapter Ten
Companion Pieces and Lighters

Three piece white porcelain set with red tea rose floral design. $30-40.

Sasha B- Germany. Cigarette holder and lighter. 1950s. $50-75 set.

Rosenthal Porcelain- cigarette holder with lighter. Pink and gray floral design. 1950s. $40-50 set.

Light orange and multi quartz ashtray with 6" tray, 3" lighter, and 4" x 5" cigarette box. $50-75 set.

White ceramic lighter with gray wheat design. Japan. $40-45.

White porcelain lighter with gray floral design. Rosenthal Porcelain. 1950s. $55-65

Black lacquered ashtray and lighter.
Wine label advertising. 1950s. $55-65.

California pottery ceramic lighter.
Japan. $20-25.

Viking art glass amber lighter. $30-35.

White ceramic lighter with shades of green
Japan. $30-35.

Silver plate horse lighter.
Occupied Japan. $40-50.

Cut crystal paperweight and lighter.
1930s-40s. $50-75.

Crystal paperweight and lighter. $30-35.

Cut crystal lighter with dimple sides.
$65-80.

Cut crystal lighter with top.
1930s. 2", $30-40.

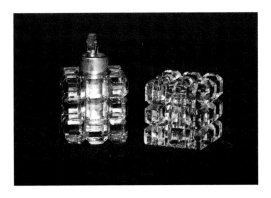

Princess House- crystal lighters. $40-45 each.

Lacquered red lighter. Japan. $20-25.

Metal lighter. Marked Germany. $40-45.

Plastic electric lighter. 1950s. $40-50.

Ruby red lighter and cigarette holder. Italian art glass. $75-100.

Hand carved Hawaiian lighter. $30-35.

Amber art glass lighter. Japan. $25-30.

Italian crystal deco lighter. $50-75.

Italian art glass amber lighter. $50-65.

Swiss made pottery lighter. 1930s. $150-185.

Italian art glass turquoise lighter. 1950s. $125-150.

Black ceramic lighter. Japan.
$30-35.

Terra-cotta " Hobo" pipe holder.
8", $25-30.

Bakelite and Jade center lighters.
Crown. $60-75.

Bibliography

Chipman, Jack. California Pottery Collectors Encyclopedia. Paducah, Ky: Collectors Books, 1992.

Schroeder's Price Guide. Paducah, Ky: Collectors Books, 1997, 1998, 1999.

Wanvig, Nancy. Collectors Guide to Ashtrays. Paducah, Ky: Collector Books, 1998.